AF534898

DETECTION OF THE COMMON FOOD ADULTERANTS

First published 1907
This edition published by Lector House in 2024

ISBN: 978-93-6138-011-2
Edition copyright © 2024 by Lector House LLP
All rights reserved under the International Copyright Conventions.

Every possible effort has been made to ensure that the information contained in this book is accurate at the time of going to press and the publisher cannot accept responsibility for any errors or omissions, however caused, in this unabridged, slightly corrected republication of the text of the first edition. No responsibility for loss or damage occasioned to any person, acting or refraining from action, as a result of the material in this publication can be accepted by the publisher. The publisher is not associated with any product or vendor mentioned in the book. The contents of this work are intended to further general scientific research, understanding and discussion, only. Readers should consult with a specialist, where appropriate.

No part of this publication may be reproduced, stored in a retrieval system, or transmitted, in any form, or by any means, electronic, mechanical, photocopying or otherwise, without the prior permission of the publisher.

Lector House LLP
Registered Office: H. No. 96, Block C, Tomar Colony,
Burari, Delhi – 110084, India
info@lectorhouse.com
www.lectorhouse.com

DETECTION OF THE COMMON FOOD ADULTERANTS

EDWIN M. BRUCE

2024

LECTOR HOUSE LLP

DETECTION OF THE COMMON FOOD ADULTERANTS

BY

EDWIN M. BRUCE

INSTRUCTOR IN CHEMISTRY,
INDIANA STATE NORMAL SCHOOL

PREFACE

BECAUSE of the recent agitation of the pure food question throughout the country, health officers, food-inspectors, and chemistry teachers and students are constantly called upon to test the purity of various foods. And this usually involves nothing more than making simple qualitative tests for adulterants. In view of the fact that there is now no text or manual devoted exclusively to the qualitative examination of foods, this little book is offered to those who are interested in this work.

Its aim is to bring together in one small book the best and simplest qualitative tests for all the common food adulterants. It contains a brief statement of the adulterants likely to be found and the reason for their use. It is hoped that it will be specially valuable to chemistry teachers in furnishing excellent supplementary work in qualitative analysis. But it is hoped that it will find its greatest usefulness in contributing something toward the great pure food reform.

It is impossible to make due mention of all the sources from which these various tests have been collected, but where possible, the author's name has been associated with the test.

TERRE HAUTE, IND.
March 25, 1907.

CONTENTS

CHAPTER III

CHAPTER IV

CHAPTER V

CHAPTER XII

PURE FOOD TESTS

CHAPTER I

DAIRY PRODUCTS

Milk—Adulterations of—Coloring matters—Annatto—Caramel—Coal-tar colors—Preservatives—Formaldehyde—Boric acid—Salicylic acid—Gelatin—Starch.

Butter—Adulterations of—Coloring matter—Preparation of sample—Annatto—Coal-tar colors—Saffron—Turmeric—Marigold—Process or renovated butter—Oleomargarine—Cottonseed oil.

MILK

Milk is adulterated by watering, removing the cream or by adding some foreign substance. Formaldehyde, boric acid or salicylic acid may be added to preserve the milk. Annatto, caramel or some coal-tar dye is added, sometimes to improve the color of the milk, and at other times to cover up traces of watering. Gelatin and starch are added for the same purpose, though they are not frequently used.

ARTIFICIAL COLORING MATTER

Annatto

Add acid sodium carbonate to a sample of the milk until it shows a slight alkaline reaction. Immerse a piece of filter-paper and leave it in for 12 or 15 hours. If annatto is present, there will be a reddish-yellow stain on the paper.

Caramel

Leach's Method.—Warm 150 cc. of the sample and add 5 cc. of acetic acid, then continue heating it nearly to the boiling point, stirring while it is being heated. Separate the curd by gathering it with the stirring rod or by pouring through a sieve. Press out all the whey from the curd and macerate the latter for several hours (10 to 12 hours) in 50 cc. of ether. It is best to do this in a tightly corked flask, shaking it frequently. If the milk was uncolored or colored with annatto the curd when thus treated will be white. If the curd is a dull brown color caramel was probably used to color the milk. Confirm its presence by shaking a portion of the curd with concentrated hydrochloric acid (sp. gr. 1.20) and gently heating. If the acid solution turns blue while the curd does not change its color, caramel was used to color the milk. (Remember that the ether-extracted curd must be brown.)

Coal Tar Colors

Lythgoe's Method.—Mix in a porcelain vessel about 15 cc. each of the sample of milk and hydrochloric acid (sp. gr. 1.20) and break up the curd into coarse lumps by shaking gently. If an azo-color was used to color the milk this curd will be pink, but the curd of normal milk will be white or yellowish.

Starch

The presence of starch in milk may be detected by heating a small quantity of the milk to boiling. When it has cooled add a drop of iodin in potassium iodid, and if starch is present there will be a blue coloration.

Gelatin

A. W. Stokes' Method.—Dissolve 1 part by weight of mercury in 2 parts of nitric acid (sp. gr. 1.42). Add 24 times this volume of water. Mix equal volumes (about 10 cc.) of this reagent and the milk or cream, shake well and add 20 cc. of water. Shake again and, after standing 5 minutes, filter. When a great quantity of gelatin is present the filtrate will be opalescent instead of perfectly clear. To a little of this filtrate in a test tube add the same volume of a saturated aqueous solution of picric acid. If much gelatin is present a yellow precipitate is produced, smaller amounts produce a cloudiness. If the filtrate is perfectly clear gelatin is absent and picric acid may be added without producing any noticeable effect.

PRESERVATIVES

Formaldehyde

Hehner's Sulfuric Acid Test.—Put 10 cc. of the suspected milk in a wide test tube and pour carefully down the side of the inclined tube about 5 cc. commercial sulfuric acid so that it forms a separate layer at the bottom. A violet coloration at the union of the two liquids indicates the presence of formaldehyde. If the commercial acid is not available, the pure acid may be used, but a few drops of ferric chloride must be added. Sometimes the charring effect of the acid makes it advisable to use the following test:

Hydrochloric Acid Test.—2 cc. of 10 per cent ferric chloride is added to one liter of commercial hydrochloric acid sp. gr. 1.2 (or any quantity in this proportion). To 10 cc. of this mixture add 10 cc. of the milk to be tested. Heat the mixture slowly nearly to the boiling point, in an evaporating dish, but agitating it all the while to prevent the curd collecting in one mass. If formaldehyde is present, there will be a violet coloration. It is said that by this test as small a quantity of formaldehyde as 1 part in 250,000 parts of milk can be detected. It is not so sensitive in sour milk.

Boric Acid

Turmeric Paper Test.—Incinerate some of the milk, and acidulate the ash with a very few drops of dilute hydrochloric acid and afterwards dissolve it in a few drops of water. Place a strip of turmeric paper in this solution for a few minutes, then remove and dry it. If boric acid either free or combined is present, the turmeric paper will be turned to a cherry-red color.

Another way of making this test.—U. S. Dep. of Agr., Div. of Chem., Bul. 65, p. 110: Make strongly alkaline with lime water, 25 grams of the milk, and evaporate to dryness on the water bath. Destroy the organic matter by igniting the residue. Dilute with 15 cc. of water and acidify with hydrochloric acid. Then add 1 cc. of the concentrated acid. Dip a piece of delicate turmeric paper in the solution; and if borax or boric acid is present, it will have a characteristic red color when dry. Ammonia changes it to a dark blue green, but the acid will restore the color.

(Turmeric paper may be prepared by dipping pieces of smooth, thin filter paper in a solution of powdered turmeric in alcohol.)

Salicylic Acid

(This is not often used as a preservative of milk.)

Leach suggests the following method for its detection.—Dissolve one gram of mercury in 2 grams of nitric acid (sp. gr. 1.42) and then add to the solution the same volume of water. Add 1 cc. of this reagent to 50 cc. of the milk to be tested, and shake and filter. The perfectly clear filtrate is shaken with ether and the ether extract evaporated to dryness. Then add a drop of ferric chlorid solution, and a violet color will be produced if salicylic acid is present.

BUTTER

Butter is often colored with annatto, saffron, turmeric, marigold or coal-tar colors. By a certain process, stale or old butter is sometimes worked over and made to appear fresh for a time. This is sold under the name of "process" or "renovated" butter. Foreign fats like cottonseed oil, sesame oil, or oleomargarine may be substituted for or added to pure butter.

COLORING MATTER

Martin's Test.—Add 2 parts of carbon bisulfid, a little at a time and with frequent shaking, to 15 parts of alcohol. Shake 25 cc. of this solution with 5 grams of the butter, and let stand for some time. The carbon bisulfid dissolves out the fatty matter and settles to the bottom. The alcohol remains on top and will dissolve out any artificial colors that may be present. If only a little coloring matter is present use more of the butter.

ANNATTO

Evaporate a portion of the extract to dryness and add sulfuric acid to the residue. If annatto is present a greenish-blue color forms. Should a pink tint result the presence of a coal-tar color is to be suspected.

COAL-TAR COLORS

These colors will dye wool or silk if pieces of the fiber are boiled in the diluted alcoholic extract, which has first been acidified with hydrochloric acid. The normal butter coloring matter will not dissolve when thus treated.

Geisler's Method.—To a few drops of the clarified fat on a porcelain surface, add a very little fullers' earth. If a pink to violet-red coloration is produced in a short time the presence of an azo-color is indicated.

SAFFRON

When saffron is present, nitric acid colors the alcoholic extract green, and hydrochloric acid colors it red.

TURMERIC

Add ammonia to the alcoholic extract, and if it turns brown it indicates the presence of turmeric.

Marigold

Add silver nitrate to the extract, and if it turns black the presence of marigold is indicated.

Process or Renovated Butter

Heat a little of the suspected butter in a spoon or dish, and if it is process butter it will sputter, but not foam much. Make the test also with some butter known to be pure and fresh.

Hess and Doolittle Test.—Melt some of the butter (say 40 grams) at about 50° C. If the butter is pure and fresh the melted fat will clear up almost as soon as it is melted, while the fat of process butter remains turbid for quite a while. After most of the curd has settled, decant as much as possible of the fat. Pour the remainder on a wet filter. Add a few drops of acetic acid to the water that runs through from the filter, and boil. If it was ordinary butter this filtrate will become milky, but if process butter a flocculent precipitate will form.

Oleomargarine

Immerse a test tube, containing some of the filtered fat, in boiling water for 2 minutes. Make a mixture of 1 part glacial acetic acid, 6 parts ether, and 4 parts alcohol. Add to 20 cc. of this mixture in a 50 cc. test tube, 1 cc. of the heated fat which may be transferred by means of a hot pipette. Stopper the tube and shake it well. Immerse in water at 15° or 16° C. Pure butter when thus treated remains clear for quite a while. There will be only a very little deposit after standing an hour, but oleomargarine gives a deposit almost immediately, and in a few minutes there will be a copious precipitate.

When the oleomargarine in butter is in about the proportion of 1 : 10, it will not separate much short of 15 minutes.

Cottonseed Oil

The presence of this oil may be detected by Halpen's test, which is given under lard, page 60.

CHAPTER II

MEATS AND EGGS

Adulterations of—Fresh and smoked—Preservatives—Potassium nitrate—Boric acid—Sulfurous acid—Salicylic acid—Benzoic acid—Canned—Preservatives (same as those of fresh and smoked meat)—Heavy metals—Coloring matter (see under sausages, etc.)—Fish, salt, dried and oysters—Preservatives—Boric acid (same under smoked and fresh meat)—Coloring matter—Aniline red and cochineal-carmine—In sausages, chopped meat, preparations and corned meat—Starch—In sausages, deviled meat and similar products—Diseased meats—Horse-flesh in sausages and in mince-meat.

Eggs—Test for age.

Meats are preserved by treating them with potassium nitrate, boric acid, sulfurous acid, salicylic acid, or benzoic acid. Cheap meat may be substituted for the more expensive. A few cases of horse meat in mince meat and sausages have been discovered. Diseased and stale meats have been found on the market. Canned meats often contain zinc, tin, and lead, and sometimes even arsenic. Aniline-red or cochineal-carmine may be added to improve the color of chopped or ground meats. Starch is sometimes added to sausage and similar meat. Fish and oysters may be preserved with boric acid or borax.

FRESH AND SMOKED PRODUCTS—PRESERVATIVES

Potassium Nitrate (Saltpeter)

Corned and smoked meats are usually preserved with saltpeter. Since smoked and cured meats are used in making potted meats, saltpeter is quite frequently found in the latter. It may be detected by the usual test for nitrates since no other nitrate is apt to be present, though one may identify the metal by the qualitative test for potassium.

To test for nitrates treat a little of the meat with 2 or 3 cc. of a 1 per cent solution of diphenylamine in strong sulfuric acid. If a nitrate is present a deep blue color forms instantly, which is not obscured by the charring effect of the acid.

Boric Acid

Pick apart into fine pieces about 25 or 50 grams of the lean meat and warm with a little water which has a few drops of hydrochloric acid in it. Soak a piece of turmeric paper in the extract, and if boric acid is present the paper will be colored rose-red when it is dry. A weak alkali turns this colored paper olive.

Another method is to burn a piece of the meat to an ash, after being treated with lime water. Make a solution of the ash and make slightly acid with hydrochloric acid. Then test with the turmeric paper with the same results as in the above method.

Sulfurous Acid

U. S. Dep. Agr., Bureau Chem., Bul. 13, Part 10: Digest 40 or 50 grams of the meat in hot water, treat with 10 cc. glacial phosphoric acid to coagulate the proteids. Strain through a cotton bag and transfer the filtrate to a short-necked flask and distil receiving the first part of the distillate in a solution of iodin. Boil, and add barium chloride. If sulfurous acid is present, it will be oxidized to sulfuric acid and precipitated as barium sulfate by the barium chlorid. More than a mere trace of the precipitate proves that some sulfite was used to preserve the meat.

Another method suggested by Kämmerer is to place the meat on paper, which has been saturated with potassium iodate moistened with dilute sulfuric acid (1 : 8); nitric oxid must not be present. If sulfurous acid is present a deep blue color forms at once. A trace of this color may form after some time with meat that is not fresh, hence this method cannot be used in examining canned meat.

Salicylic Acid

Heat 50 grams of the meat in 50 cc. of water. Add 10 cc. of a strong solution of glacial phosphoric acid and strain through a cotton bag. Extract the filtrate with a little ether (about 50 cc.) in a separatory funnel. Let the ether evaporate spontaneously. Take up the residue with 3 cc. of water, and add one or two drops of a one-half per cent solution of ferric chlorid. If salicylic acid is present the mixture will be purple or violet.

Leach makes the same test by slightly acidifying a portion of the lean meat, then extracting with ether, and evaporating to dryness and testing the residue with a drop of ferric chlorid solution. A deep violet coloration is produced if salicylic acid is present.

Benzoic Acid

Mohler's Method.—Prepare a sample as in the test for salicylic acid by heating 50 grams of the meat in 50 cc. of water. Add 10 cc. of a concentrated solution of glacial phosphoric acid, and strain through a cotton bag. Neutralize with sodium hydrate and evaporate to dryness or to a small volume. After treating with 3 cc. of concentrated sulfuric acid, heat till white fumes appear. Add 4 or 5 crystals of potassium nitrate and continue heating until the solution is colorless or nearly so. When cool dilute with water, add an excess of ammonia, and place in a narrow vessel like a test tube. Add one or two drops of ammonium sulfid carefully so that the liquids do not mix. If the surface of the liquid immediately becomes red, benzoic acid is present.

If this test is not carefully performed, it is worthless, as other substances give similar results.

Confirm its presence by neutralizing the aqueous solution of the extracted benzoic acid with sodium hydroxid; concentrate to a very small volume. Acidify with sulfuric acid. A white flocculent precipitate shows the presence of considerable benzoic acid.

CANNED MEAT

If in preparing canned meat only fresh meat is used, there is little need for the use of preservatives, but as considerable smoked and cured meat is thus used, preservatives may find their way into canned meat.

The same preservatives should be looked for as in fresh and smoked meat and the same test made for each.

HEAVY METALS

A. H. Allen's Method.—About 25 grams of the substance is mixed slowly with enough strong, pure sulfuric acid to just moisten the mass, avoiding an excess. Heat on a water-bath for a short time, then raise the temperature gradually, and maintain till the chlorids seem to be decomposed. It must not be hot enough, however, to volatilize the sulfuric acid. Then add 1 cc. of strong nitric acid and heat till red fumes are given off. Freshly ignited magnesia in the proportion of 0.5 gram for each cc. of sulfuric and nitric acid previously used is now stirred into the mass and the whole ignited at a dull red heat. This is best done in a gas-muffle. When cool, moisten the ash with nitric acid and gently re-ignite, repeating this treatment till the carbon is entirely consumed. Treat the residue with 8 or 10 drops of strong sulfuric acid, heat till fumes are given off, cool, boil with water, dilute to about 100 cc. and saturate with hydrogen sulfid, filter, examine as follows:

Zinc and iron may be in solution. Add bromine water to destroy hydrogen sulfid and to oxidize the iron, boil and add ammonium hydrate in excess, boil again and filter.		Lead, tin, copper, and calcium, if present, will be in the precipitate and residue. Fuse the mass in a porcelain crucible for at least ten minutes with 2 grams each of potassium and sodium carbonates and half as much sulfur. After cooling, boil with water and filter.	
The precipitate will contain the iron and the phosphates.	Filtrate, when blue, proves the presence of nickel.	Residue. Add hydrochloric acid and boil as long as hydrogen sulfid is given off. A few drops of bromine water will complete the oxidation of the copper sulfid. Filter, and add ammonium hydroxid in excess to the filtrate.	The filtrate may contain tin. Acidify with acetic acid, and if tin is present a yellow precipitate of stannic sulfid will form.

		When the filtrate is blue, it indicates the presence of copper. Acidify the filtrate with acetic acid and test for lead by adding potassium chromate, a yellow precipitate being formed when it is present.	
	I. Heat to boiling and add potassium ferrocyanid. A white precipitate or turbidity indicates *zinc*.		

FISH SALT DRIED AND OYSTERS

This kind of meat is often preserved with boric acid and borax. They may be detected by the method given under fresh and smoked meat.

COLORING MATTER

Sausages and other chopped meat preparations, together with corned meat that has been cured without saltpeter, are often treated with artificial coloring matter.

Aniline red and cochineal carmine are usually employed for this purpose. The former may be detected, according to Allen, by picking the meat apart and treating it with methylated spirit, strain or filter the extract and take up with water. Then a piece of white wool (nun's veiling will do) is immersed in the boiling liquid and, if it is dyed red, rosaniline is present.

Cochineal-carmine may be detected by the method used by Klinger and Bujard. Cut up fine about 20 grams of the meat and heat in a water-bath with water and glycerine mixed in equal parts. If the above coloring matter is present the liquid will become quite red in color, if not present a slight yellow color results from this treatment. If a spectroscope is available this dye is easily recognized.

STARCH

In Sausage, Deviled Meat, and Similar Products

Cracker and bread crumbs are often added to these preparations and their presence is best detected by examining the amount of starch present. Do this by boiling some of the sample in water, and when cool adding a drop or two of iodin reagent. The usual blue color is produced if much starch is present. If there is only a little starch, it may be necessary to examine it under the microscope to determine whether the starch is from the pepper and other spices used or from some cereal.

DISEASED MEAT

The following method is recommended by Ebers.—Hold a small piece of the suspected meat over a mixture of 1 cc. hydrochloric acid, 3 cc. alcohol, and 1 cc. of ether. The formation of ammonium chlorid fumes shows that decomposition has begun. Do not mistake the fumes of the acid for those of ammonium chlorid.

HORSE FLESH

In Sausage and Mince Meat

This sophistication is not common in this country. Horse flesh is detected by testing for glycogen, which is present in it in larger quantities than in other meats.

Courley & Coremon's Test.—Boil 50 grams of the meat for a half hour with water, strain, and to a portion of the filtrate add a few drops of potassium iodid-iodin solution (potassium iodid 0.4 gram; iodin 0.1 gram; water 20 cc.). If considerable horse meat is present the glycogen will color the liquid dark brown, which disappears on heating, but returns on cooling.

EGGS

It sometimes happens that one wishes to know the age of eggs without opening them.

Delarne's Test.—Place the egg in a 10 per cent solution of common salt. Perfectly fresh eggs sink to the bottom. Those remaining immersed, but suspended in the liquid, are at least three days old, while those rising to the surface and floating are more than five days old. The older the egg, the higher it floats and the more it will stand on end. This test is not applicable to eggs that have been preserved.

Hold the egg between a bright light and the eye, and if the air chamber is small, and no dark spots but instead a rather uniform rose-colored tint is seen, the egg is fresh. If the contents appear cloudy and the air chamber larger, the egg is not fresh. The darker the contents of the egg the older it is.

CHAPTER III

CEREAL PRODUCTS

Flour—Adulteration of—Alum—Copper sulfate—Substituted flours—General test—Corn meal in wheat flour—Wheat flour in rye flour—Ergot in rye flour.

Bread—Adulterations of—Alum—Copper sulfate.

Ginger Cake—Adulterations of—Stannous chlorid.

FLOUR

SOMETIMES a cheaper or inferior grade of flour is substituted for one of higher quality, and even a different kind of flour may be substituted, as corn meal in wheat flour, or wheat in rye flour. Alum may be added by millers to cover up traces of bad flour, and by bakers to make the bread white when a bad or cheap flour is being used. Copper sulfate also may be added to improve the appearance. Occasionally rye flour is made from rye upon which ergot has developed. Stannous chlorid and potassium carbonate are added to ginger cake to give the same color to the product made of molasses and a poor grade of flour as that made from good flour and honey.

ALUM

Wynther Blyth Method.—Add a little water to the sample and macerate. Soak pieces of gelatin in the solution and leave for a half day, remove the gelatin and dip the pieces in a mixture of equal volumes of a fresh tincture of logwood and a saturated solution of ammonium carbonate. The gelatin strips will turn blue if alum is present.

Bell & Carter Method.—Make a fresh 5 per cent tincture of logwood in methyl alcohol. Dampen about 10 grams of the flour with water and add 1 cc. of the logwood tincture and the same quantity of a saturated solution of ammonium carbonate. Pure flour gives a pinkish color which fades to buff or brown. The presence of alum produces a lavender or bluish tint which becomes more distinct as it dries.

COPPER SULFATE

This adulterant may be detected in either flour or bread, by soaking the flour or bread in a dilute solution of potassium ferrocyanid acidulated with acetic acid. If copper be present a purplish or reddish-brown coloration will be produced.

SUBSTITUTED FLOURS

Vogel's Method.—Make a mixture of alcohol (70 per cent), 95 parts, hydrochloric acid 5 parts. Treat a sample of the flour in a test tube with this reagent. Shake well. Heat to boiling and allow to settle. A colorless fluid shows the flour to be pure, a straw-colored tint indicates the presence of gruffs with bran, an orange-yellow proves the presence of corn-cockle flour, a flesh-colored liquid indicates the presence of ergot, while a green color indicates buckwheat flour.

Corn Meal in Wheat Flour

Kraemer claims to be able to detect as small amount as 5 per cent of maize in wheat flour, by the following test.—Mix a gram of the flour with 15 cc. of good glycerin, and heat to boiling for a short time. If corn meal is present, there will be an odor like that of pop corn.

Wheat in Rye Flour

Kleeburg recommends the following test.—A little of the flour is mixed on a piece of common window glass or microscope slide, with sufficient water (at 45° C.) to float the flour particles. Spread the mixture out over the glass, and press another glass down upon it. When wheat flour is present, white spots will be seen, and if the glasses are slid upon each other the spots will pull out into threads, and the thicker and longer they are the more wheat flour there is present.

Ergot in Rye Flour

Boettger gives the following chemical test for ergot.—Heat 10 to 15 minutes with an equal quantity of ether, adding a few crystals of oxalic acid. When ergot is present a reddish color develops.

Another Method.—Bul. 51, Bureau of Chem.

Digest 20 grams of the suspected flour, with boiling alcohol, till no more color is extracted. Add 1 cc. of sulfuric acid (1 : 3), and if ergot is present the solution will be colored red.

BREAD

Alum

Moisten a piece of the bread with water, and then with a logwood solution (5 grams logwood digested in 100 cc. of alcohol). If alum is present the bread will become lavender blue in two or three hours. Pure bread would have a red-brown tint. To prove the presence of alum, the blue color must be permanent at the temperature of boiling water. (The logwood used in this test must be pure.)

Blyth's Test.—Macerate 150 grams of the sample for 45 or 50 hours in a couple liters of water; after straining through muslin, evaporate to a small volume over a low flame. Immerse a strip of gelatin in this liquid, and then in a logwood solution (same as in last test), and if alum is present it will acquire the lavender color.

If the bread in either of these tests is sour, the following modification (Vanderplanken) must be made. Reduce 15 grams of the sample to a paste with water and some pure chlorid of sodium, adding 10 drops of a fresh alcoholic solution of logwood, after which add 5 grams of pure potassium carbonate. Mix well, and after washing with 100 cc. of water into a vessel allow to settle. If alum is present the liquid will soon become reddish-violet, and if not present it will be blue.

COPPER SULFATE

See Test for Copper Sulfate in Flour.

GINGER CAKE

Tin may be detected by the method for heavy metals under meat.

CHAPTER IV

LEAVENING MATERIALS

Baking Powders—Adulterations of—Tartaric acid (free or combined)—Tartaric acid (free)—Sulfates (calcium, etc.)—Gypsum—Ammonium salts—Alum.

Cream of Tartar—Adulterations of—Tartaric acid (free or combined)—Aluminium salts—Ammonia—Earthy materials.

BAKING POWDERS

Baking powders consist of bicarbonate of soda and an acidifying agent as acid potassium tartrate, acid calcium phosphate, tartaric acid or alum. Some powders contain both acid calcium phosphate and alum. The kind of powder is determined by testing for these. Gypsum has been added to baking powders to increase the weight.

TARTARIC ACID

Free or Combined

Wolff's Method.—If no starch is present, mix a little of the powder with some dry resorcin. Add a few drops of sulfuric acid and heat gently. A rose-red color forms if tartaric acid or tartrates are present. The color should disappear when diluted with water. When starch is present, mix well by shaking about 5 grams of the powder with 250 cc. of cold water. Let the insoluble matter settle and pour the liquid upon a filter. Evaporate the filtrate to dryness, treat the powdered residue with a few drops of a 1 per cent solution of resorcin. Add 3 cc. of strong sulfuric acid, heat slowly. A rose-red color forms if tartaric acid is present. The color should be destroyed on the addition of water. This test is applicable in the presence of phosphates and the acid may be free or combined.

TARTARIC ACID

Free

Make an absolute alcoholic extract of 5 grams of the powder and evaporate the alcohol. Add sufficient dilute ammonia to dissolve the residue, place in a test tube and drop in a crystal or two of silver nitrate. Heat gently, and a silver mirror will form if tartaric acid is present.

SULFATES

Calcium, etc.

Boil a portion of the sample gently with strong hydrochloric acid, add barium chlorid. A white precipitate of barium sulfate will form if sulfuric acid is present.

GYPSUM

Calcium Sulfate

Ash a portion of the sample and make the usual qualitative tests for calcium sulfate.

AMMONIUM SALTS

Extract a few grams of the sample with cold water, boil the extract with sodium hydroxid and place a piece of moist red litmus paper in the steam. It will be colored blue if ammonia is present.

ALUM

Reduce to an ash about 2 grams of the powder in a platinum dish. Extract with boiling water, add ammonium chlorid solution to the filtrate until a distinct odor of ammonia is given off.

If a flocculent precipitate forms it indicates the presence of alum.

This test for alum is applicable in the presence of phosphates.

Mrs. Richards.—Cover some logwood chips (they must be pure) with water and bring to a boil. Repeat this four times, saving only the last decoction. Shake some of the sample (a couple of teaspoonfuls) in a beaker half full of water. When it ceases effervescing, strongly acidify with acetic acid. Add a few drops of the logwood extract, and if alum is present a bluish-red color will appear.

CREAM OF TARTAR

Cream of tartar is bitartrate of potassium and is obtained from the lees deposited in wine casks. If gypsum has been used to clarify the wine, it will be present in the cream of tartar as calcium tartrate.

Other adulterants of cream of tartar are acid calcium phosphate, starch, gypsum, and alum.

TARTARIC ACID

Free or Combined

If the sample is known to be free from starch the following test may be made:

Mix a bit of the powder with a small quantity of dry resorcin and add a few drops of concentrated sulfuric acid. Heat slowly, and if a rose-red color forms, which disappears when diluted with water, there is present either tartaric acid or a tartrate.

When the sample contains starch, shake about 4 or 5 grams of it a number of times with 250 cc. of cold water in a large flask. Pour on a filter after the insoluble material has settled and evaporate the filtrate to dryness. The residue is to be tested for tartaric acid and tartrates, the same as when starch was absent.

ALUMINIUM SALTS

Mix equal quantities (about 1 gram) of the sample and sodium carbonate and burn to an ash. Extract with boiling water and filter. Add to this filtrate enough ammonium chlorid solution to cause a distinct evolution of ammonia. The formation of a flocculent precipitate shows the presence of aluminium. This test may be used when phosphates are present.

AMMONIA

Present in the Form of Ammonia Alum or Ammonium Carbonate

Make a cold water extract of the powder and boil it with sodium hydroxid. Test the steam with moist red litmus paper.

EARTHY MATERIALS

Treat the sample with warm potassium hydroxid. A residue indicates some earthy material.

CHAPTER V

CANNED AND BOTTLED VEGETABLES

Adulterations of—Preservatives—Preparation of sample—Formaldehyde—Sulfurous acid and the sulfites—Salicylic acid—Saccharin—Benzoic acid—Coloring matter—Cochineal—Coal-tar dyes—Copper salts—In green pickles, beans, peas, etc.—Turmeric—In mixed pickles—Heavy metals (other than copper, same as under meats)—Soaked vegetables—Peas, beans and corn—Alum—In pickles—Examination of the can or box.

No class of foods on the market has less need for antiseptics than canned goods, yet their use is rather common. Products thus treated are easier canned and are not so apt to spoil. The chemicals used as preservatives are sulfurous acid, and the sulfites, salicylic acid and saccharin, benzoic acid, and sometimes formaldehyde. Sulfurous acid is used to bleach such foods as canned corn. Saccharin possesses some antiseptic properties, but its main use is as a sweetener. Alum is used to make pickles hard and crisp.

Some canned or bottled goods, as tomato-catsup, is colored with cochineal or coal-tar dyes. Green pickles, beans, peas, and such vegetables are colored by copper salts or are cooked in copper vessels, with the addition of acetic acid, hence the beautiful green color. Turmeric is sometimes used to color mixed pickles.

The heavy metals as lead, zinc, and tin are generally present in canned goods, the amount varying with the corrosive power of the vegetable.

When there is a year of scarcity in corn, peas, beans, and such vegetables, the dried product is often soaked and canned. Some of this goods is sold for the regular green vegetable, while some may be properly marked "Soaked Goods."

PRESERVATIVES

It is best to make a systematic examination for the different preservatives. The sample may be prepared by mixing 50 grams of the pulped material with sufficient water in a 250 cc. graduated flask. Add phosphoric acid till distinctly acid in reaction. Fill to the mark with water. Place in a distilling flask, and distil in a linseed oil or a paraffin bath till 30 cc. have been collected. Save this distillate for the following tests.

FORMALDEHYDE

To 5 cc. of the above distillate in a test tube, add 2 or 3 drops of a 1 per cent aqueous solution of phenol and mix well. Incline the tube and carefully pour down the side 5 cc. of concentrated commercial sulfuric acid so that the two liquids do not mix. If formaldehyde is present there will be a crimson zone at the plane of union of the solutions. This coloration takes place when the formaldehyde is present in the proportion of 1 part in 100,000 parts. When there is a greater quantity of formaldehyde present a white turbidity or a light-colored precipitate forms above the coloring.

Phenylhydrazine Hydrochloric Test.—Dissolve 2 grams of phenylhydrazine hydrochlorid and 3 grams of sodium acetate in 20 cc. of water. Add 2 to 4 drops of this reagent and the same number of drops of sulfuric acid to 1 or 2 cc. of the above distillate, to be examined in a test tube. A green coloration is produced when formaldehyde is present.

Hydrochloric Acid Test.—Add 5 cc. of the distillate to be tested to about 5 cc. of milk known to be pure, and about 10 cc. of concentrated hydrochloric acid (sp. gr. 1.2) which contains 1 cc. of a 10 per cent ferric chlorid solution to each 500 cc. of the acid. Heat slowly to 80° or 90° C.

over the free flame, agitating it at the same time to break up the curd. A violet coloration indicates formaldehyde.

Sulfurous Acid and the Sulfites

Free sulfurous acid is not largely used as a food preservative, though its salts are quite commonly employed.

Detection.—Mix 150 grams of the finely ground sample with enough water to make a thin paste. Acidify with phosphoric acid and distil till 25 cc. have been collected. (The delivery tube of the condenser should dip below the surface of a little water.) Treat the distillate with a few drops of bromine water and boil for a short time. If a precipitate forms on the addition of barium chlorid the presence of sulfurous acid is indicated.

Salicylic Acid

Acidify 50 cc. of the sample with sulfuric acid, and shake vigorously with 50 cc. of a mixture of equal parts of ether and petroleum spirit. When the liquids have separated, draw off as much as possible of the solvent and filter. If an emulsion forms use a centrifugal machine, and evaporate with a small flame. If needle-shaped crystals form, salicylic acid is present. Add a few drops of water and a drop of very dilute ferric chlorid solution in such a way that the solutions will come together slowly. The presence of salicylic acid gives a purple or violet color.

Saccharin

This is used quite extensively as a sweetening agent in canned sweet corn, and other similar products.

Macerate about 20 grams of the sample after mixing with 30 to 40 cc. of water and strain through muslin. Acidify with 1 or 2 cc. of sulfuric acid (1 to 3) and extract with ether. (If an emulsion forms, use a centrifugal machine.) Separate the ether layer and let the ether evaporate spontaneously and use the residue in the following tests:

Take up a part of the residue with water and taste. If it is very sweet saccharin is present. Confirm by the following:

Schmidt's Test.—Add about 1 gram of sodium hydroxide to another part of the residue, and heat in an air-oven or oil bath, for half an hour at about 250° C., to convert the saccharin into salicylic acid. After it has cooled, acidify with sulfuric acid, extract and test for salicylic acid with 2 or 3 drops of ferric chlorid solution, letting the solutions come together slowly. A purple or violet coloration proves the presence of salicylic acid, which in turn indicates the presence of saccharin. This test cannot be used if salicylic acid was used as a preservative in the original product. A test for the acid should first be made.

Bornstein's Test.—Heat the remainder of the above ether residue with resorcin and a very little sulfuric acid till it begins to swell. (It is best to do this heating in a test-tube.) Let cool till the action stops, heat again and repeat the operation several times. After cooling the last time, dilute with water and add sodium hydrate till neutral. If saccharin is present, there will be a red-green fluorescence.

BENZOIC ACID

Acidify 50 cc. of the sample with sulfuric acid and shake vigorously with 50 cc. of a mixture of equal parts of ether and petroleum spirit. Let the liquids separate, then draw off as much as possible of the solvent and filter. (Use a centrifugal machine if an emulsion forms.) Separate the extract into 2 parts and evaporate each to dryness over a small flame and make the following tests:

Ferric Chlorid Test.—Dissolve one of these residues in ammonia, and evaporate to dryness on a water-bath. Take up the residue with warm water, filter, and collect the filtrate in a small test tube. Add a drop of ferric chlorid solution, and if benzoic acid is present a characteristic flesh or brownish colored precipitate of ferric benzoate forms. Sometimes in such products as sweet pickles, a basic ferric acetate precipitate comes down and the following test had better be applied.

Peter's Method.—Take about 0.1 gram of the second part of the above ether residue, place in a large test tube (about 50 cc.) and dissolve in 5 to 8 cc. of concentrated sulfuric acid. Add from 0.5 to 0.8 gram of barium peroxide, a little at a time. Shake each time and cool in water if necessary. This should produce a permanent froth on the sulfuric acid. Let stand 25 or 30 minutes, then fill the tube three fourths full of water, shake and cool rapidly to the temperature of the room, and filter off the barium sulfate. Extract with chloroform or ether. Remove the extract and test it for salicylic acid with dilute ferric chlorid. (See first test under salicylic acid.) In this method salicylic acid must first be proven absent.

Mohler's Test.—Treat the remainder of the second part of the above ether residue with 2 or 3 cc. of concentrated sulfuric acid. Heat till white fumes appear. Add a few crystals of potassium nitrate and when cool dilute with water. Add an excess of ammonia, then a drop or two of ammonium sulfid. If a red color appears immediately on the surface, it shows the presence of benzoic acid.

COLORING MATTER IN CATSUPS AND TOMATOES

COCHINEAL

Girard and Dupre Test.—Shake well a portion of the sample with water and filter, acidify with hydrochloric acid, then extract with amyl alcohol, and if cochineal is present the extract will be colored yellow or orange, the particular shade depending on the amount of cochineal present. Remove the amyl alcohol and wash with water until it is neutral. To half of this, add a very dilute solution of uranium acetate, drop by drop, and shaking well after the addition of each drop. Cochineal, if present, will produce a characteristic emerald-green color.

Confirm by adding a drop or two of ammonia to the second half of the amyl alcohol extract and a violet coloration will be produced if cochineal is present.

Coal-Tar Coloring Matter

Sostegni and Carpentieri Test.—Free from grease a piece of woolen cloth (nun's veiling will do) by boiling first in very dilute caustic soda solution and then in water. Acidify a portion of the sample with 2 to 4 cc. of 10 per cent solution of hydrochloric acid and filter. Strips of the cleansed cloth are boiled in this filtrate for 5 or 10 minutes, then removed, washed in water and boiled with very dilute hydrochloric acid solution. Wash out the acid and dissolve the color from the cloth by boiling in a solution of ammonium hydroxid (1 to 50). (The time required will depend upon the dye present.) Remove the cloth from the solution and acidify the latter with hydrochloric acid and another piece of the cleansed cloth is immersed and again boiled. This second dyeing fixes only coal-tar colors on the cloth, hence, no fear of mistaking them for the natural color of the vegetable.

IN GREEN PICKLES, BEANS, PEAS, ETC.

Copper Salts

Burn 20 grams of the sample to an ash and wet the ash with concentrated nitric acid, dilute with water and boil. Add ammonia till strongly alkaline and filter. If the filtrate is blue, copper is present.

Confirm by acidifying the filtrate with acetic acid and adding potassium ferrocyanid. A red or brownish precipitate or coloration proves the presence of copper. The test for other heavy metals may be made by the general method given under meats.

IN MIXED PICKLES

Turmeric

Shake with alcohol to extract the color. Soak a piece of filter paper in the extract and dry in an air oven at 100° C. Wet the filter paper with a weak solution of boric acid to which a very little hydrochloric acid has been added. If turmeric is present, a cherry-red color will appear when the filter paper is dry.

"SOAKED" VEGETABLES

Peas, Beans, and Corn

There is really no chemical test for this class of foods. Certain helpful directions given in Bul. 65, p. 54, of the Bureau of Chem., will assist in identifying such goods. All or nearly all of the green color of peas and beans is destroyed by the process of "soaking." They have the appearance of the well-matured product, and are firm and mealy with well-formed cotyledons. The process of soaking starts the growth of the caulicle of the pea. The kernel of corn is plump and hard and does not have the milky consistency of the immatured product. The characteristic succulence of the green pea, bean, and corn is absent in the soaked product.

Alum in Pickles

This is sometimes added to the pickling solution to produce hardness and crispness.

Burn to ash a sample of the pickles, and, if they are free from copper, fuse in a platinum dish with sodium carbonate. Extract with boiling water, and after filtering add ammonium chlorid solution. If alum is present, a flocculent precipitate will form.

Examination of the Can or Box in which Vegetables are Sealed

Generally when the ends of a can are convex, instead of plane or concave, it is spoiled. In the souring of canned sweet corn, it is exceptional that the ends are forced outward.

Strike the can and the spoiled cans will give a dull sound while the good ones will give a distinct tone. Some practice will be necessary to use this test.

One can judge of the amount of tin dissolved by the corrosion of the inside of the can.

Reject cans that show much rust around the cap on the inside of the head.

If more than one hole is found soldered in the cap, reject the can. Cans of salmon are the only exception that has come to the author's notice. A second hole, in general, indicates that decomposition had set in and the can had been punctured and resealed.

CHAPTER VI

FRUITS AND FRUIT PRODUCTS

Adulterations of—Preservatives—Preparation of sample—Salicylic acid—Benzoic acid—Saccharin—Coloring matter—Coal-tar dyes—Cochineal—Acid magenta—Apple juice in jellies made from small fruits—Detection (see test for starch)—Starch—In jellies, jams and such products—Gelatin—In jellies—Agar agar—Heavy metals—Arsenic.

SALICYLIC acid, benzoic acid, and saccharin are used to preserve fruits. The last is also added as a sweetener instead of sugar. Many fruit products lose their color with age, and to give them a lasting color they are treated with a coal-tar dye, cochineal, acid magenta, or caramel.

A very small per cent of the jams and jellies sold are strictly pure. These cheap products are made up principally of apple juice and commercial glucose; artificial essences are added to imitate the real flavor.

In cheap jellies made of apple juice and glucose syrup, a "coagulator" is used; usually sulfuric acid and alum, also citric and tartaric acids may be used for this purpose.

Starch, gelatin, and agar are used as gelatinizing agents.

Fruits put up in tin cans may dissolve the heavy metals as tin, zinc, lead, and even arsenic.

PRESERVATIVES

Preparation of the Sample.—Dissolve 25 or 30 grams of the sample in water which has been acidified with sulfuric acid (1 to 3), then extract with ether, and remove the ether layer and let it evaporate spontaneously. The residue may contain salicylic acid, benzoic acid, or saccharin. Take up with a little water and make the following tests:

Salicylic Acid

Place a few drops of this extract in a test tube and add a drop or two of a 0.5 per cent solution of ferric chlorid. If salicylic acid is present, there will be a purple coloration.

Benzoic Acid

Mohler's Test.—Add 2 to 3 cc. of strong sulfuric acid to a second portion of the above ether extract and heat until white fumes appear. Then add a few crystals of potassium nitrate and heat again. Continue adding the nitrate and heating till the solution is colorless or only a very light yellow. Dilute with about 5 cc. of water when cool, neutralize with ammonia. It should be filtered when not clear or when crystals of ammonium or potassium sulfate are formed. Add a few drops of ammonium sulfid to the filtrate in such a way as to prevent the mixing of the liquids. The sulfid will be on top. If a bright cherry-red color forms where the two liquids meet, either benzoic acid or saccharin is present. Distil and the benzoic acid will pass over, extract the distillate in the usual way and apply the above test to it for benzoic acid.

Saccharin

Taste a third portion of the ether extract. A very sweet taste indicates saccharin. A further test can be made by adding 1 or 2 grams of sodium hydroxid to the rest of the ether extract and heating a half hour in an oil bath at 250° C. Dissolve in water when cool, acidify with dilute sulfuric acid and extract with ether. The saccharin will have been converted into salicylic acid, which may be identified by the usual test for that acid. This test presupposes the absence of salicylic acid in the original material.

COLORING MATTER

Coal-Tar Dyes

To attempt to identify the particular dye used in every case would be quite beyond the object of this set of simple tests. A general test showing the presence of a coal-tar dye is probably all that is usually desired.

Sostegni and Carpentieri Test.—Such a test may be made by dissolving 15 grams of the fruit product in 100 cc. of water, filtering and acidifying with a small quantity of a 10 per cent solution of hydrochloric acid and again filtering. Place in the filtrate strips of white woolen cloth (nun's veiling will do) which have been freed from grease by boiling first in very dilute caustic soda solution, then in water, and boil for 5 to 10 minutes. Remove the cloth and wash it in water, then boil in very dilute hydrochloric acid. Stir the cloth in water to remove the acid and dissolve the color by boiling in a solution of ammonium hydroxid (1 to 50). The time required will depend upon the particular dye used. Remove the cloth from the solution and acidify the latter with hydrochloric acid, a slight excess is better, and another piece of the cleansed cloth is immersed and again boiled. Nothing but coal-tar dyes will color in this second dyeing.

Cochineal

Girard and Dupre Test.—See tests for cochineal under "Catsups and tomatoes."

Acid Magenta

Girard and Dupre.—Make about 100 cc. solution of the fruit, filter, and neutralize with potassium hydroxid (strength 5 to 100); about 2 cc. will be needed. Add 4 cc. of mercuric acetate solution (1 to 10), shake and filter. By this treatment the filtrate should be colorless and slightly alkaline.

Add sulfuric acid till there is a slight excess. A colorless solution indicates the absence of acid magenta, while a light violet-red shows its presence, providing the amyl-alcohol extract showed no other dye to be present.

Caramel

Amthor's Test.—10 cc. of a solution of the fruit is put into a deep, narrow glass (a bottle may be used). Add 30 to 50 cc. of paraldehyde, to be gaged by the intensity of the coloring. Then add a sufficient quantity of absolute alcohol to make the solutions mix. If caramel is present, a brownish-yellow to dark-brown precipitate will be formed, decant, wash the precipitate once with absolute alcohol, dissolve in a little hot water and filter. The shade of color is proportional to the amount of caramel present.

To verify the test, pour the colored fluid into a freshly prepared solution of phenylhydrazin (2 parts phenylhydrazin-hydrochlorid, 3 parts sodium acetate, and 20 parts water). Much caramel produces a dark-brown precipitate in the cold, and is hastened by slightly heating. A very small amount of caramel will require several hours to precipitate.

APPLE-JUICE IN JELLIES MADE OF SMALL FRUITS

Very often cider is added to other fruit juices to give them the proper consistency in jellies, jams, and marmalades.

Its presence may some times be determined by making the usual starch test. A large quantity of starch is normally present in apples, but is less as they ripen, and finally disappears in the ripened fruit. There is no starch, or only a mere trace, in small fruits even when green. It is readily seen that if the juice is taken from green apples that there will be starch found in the artificial jelly or jam, though its absence does not prove the absence of cider.

Make the starch test as follows:

STARCH

In Jellies, Jams, and Such Products

Make a solution of the jelly or jam and destroy the color by heating nearly to the boiling point and adding dilute (1 : 3) sulfuric acid and potassium permanganate until the color is destroyed. This treatment does not affect the starch, and when cool add iodin, preferably potassium iodid-iodin (potassium iodid, 0.4 gram; iodin, 0.1 gram; water, 20 cc.). If a great quantity of starch is present an almost black precipitate will be formed. Smaller amounts give the usual blue color.

Whenever starch is found to be present, it is best to make a microscopical examination in the case of jams and marmalades. If the starch is normally present the grains will be seen within the cell walls after the iodin treatment.

Starch is nearly always present in the apple and some other fruits, so unless it is present in jelly and such products in considerable quantity it is not likely that it was added.

GELATIN

In Jellies

Henzold Test.—Add water to some of the jelly and boil for a short time, filter and treat the filtrate with an excess of a 10 per cent solution of potassium bichromate and boil again. After cooling add 2 or 3 drops of concentrated sulfuric acid. A white flocculent precipitate forms if gelatin is present, and it gradually collects in a lump at the bottom.

E. Beckmann's Method.—Treat the jelly with 95 per cent alcohol and wash the precipitate with alcohol to free it from the sugar, then drive off the alcohol by heating. Add a very little water to the residue and neutralize the extract with calcium carbonate. Then add formalin and evaporate to dryness. By this treatment gelatin is rendered insoluble.

Pure fruit jellies have only 1 to 2 per cent of insoluble precipitate, while those jellies in which gelatin is used have 70 to 86 per cent of insoluble precipitate.

AGAR AGAR

Boil the sample with 5 per cent sulfuric acid. Add a crystal or two of potassium permanganate, and wait till it settles, and examine the sediment for diatoms with a microscope. Their presence shows the use of agar.

HEAVY METALS

Tin, Zinc, Lead, and Copper

A. H. Allen's Method.—(See test for heavy metals under canned meat.)

ARSENIC

Marsh's Test.—Fit a 100 cc. flask with a two-holed rubber stopper, through which passes a long-stemmed separatory funnel reaching nearly to the bottom, and a delivery tube which connects with a bulb tube containing a little acetate of lead solution. This in turn is connected with a calcium chlorid tube and this with a small, hard glass tube, 15 or 20 cm. long, not over 0.5 cm. bore, and drawn to small size in the middle. The large part next the chlorid tube is protected by fine wire gauze which extends to within a half inch of the constricted part. Two burners may be so placed as to heat the gauze. The flask should be placed in water and the bulb tube may be. Four grams of arsenic-free zinc, and 40 cc. of dilute pure sulfuric acid (1 to 8) are placed in the flask. Let the hydrogen flow at least a quarter of an hour, then heat the gauze for 15 or 20 minutes. There should be no deposit in the tube. Now, char a portion of the sample, dissolve in water and pour into the separatory funnel, letting it run slowly into the flask. A dark deposit in the glass tube shows that arsenic is present, but if after an hour no darkening takes place it is quite safe to say that no arsenic is present in the fruit.

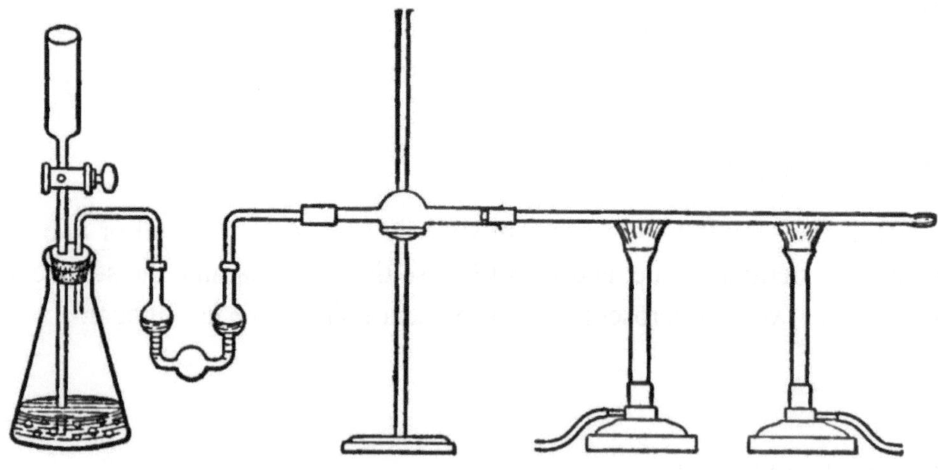

Apparatus for Marsh Test

Gutzeit's Test.—Place a gram of pure zinc, 5 cc. of dilute sulfuric acid (6 per cent) and about 1 cc. of a solution of the sample in a deep test tube. Cover the tube with three thicknesses of filter paper, fitted tightly over the mouth of the tube. Place on the upper paper a drop of strong silver nitrate solution. Place the tube in a dark place and leave for 10 minutes. If a bright yellow stain forms on the filter paper, and turns black or brown when water is added to it, arsenic is present. Unless one is certain of the purity of the reagents used it is advisable to make a blank test, using distilled water instead of the fruit.

Oxidize all sulfids to sulfates before applying the above test. To find out whether they are present or not, substitute lead acetate for the silver nitrate on the filter paper. To avoid some of these difficulties treat according to the following method:

Preparation of the sample according to Leach.—If possible reduce the sample to a dry char by treatment with concentrated nitric and sulfuric acids so that it may be powdered in a mortar. Dissolve out the arsenic by repeated treatment with boiling water. Save this extract, and, when cool, filter and submit to Marsh's test which is given above.

In case the sample is too much of a solid to get the arsenic out by the above treatment, it may be prepared according to the directions of Chittenden and Donaldson: Heat 100 grams of the macerated sample with 23 cc. of pure, strong nitric acid to a temperature of 150° C. or 160° C. Assist the action by stirring occasionally. When the fruit becomes a deep yellow or orange color, remove the heat and add 3 cc. of pure, strong sulfuric acid. It should be stirred while nitrous fumes are passing off. Heat again to about 180° C., and before it cools add, drop by drop, 8 cc. of pure concentrated nitric acid. It should be stirred constantly while the acid is being added. Heat at 200° C. till sulfuric acid fumes begin to come off and only a dry mass remains. Powder the mass and exhaust it with boiling water, filter and test the solution when cold with Marsh's test.

CHAPTER VII

FLAVORING EXTRACTS

Lemon Extract—Lemon oil—Citral—Oil of citronella—Tartaric or citric acid—Methyl alcohol—Coloring matter—Turmeric—Coal-tar colors.

Vanilla Extract—Adulterations of—Preliminary test—Alkali—Foreign resins—Caramel—Tannin—Coumarin.

LEMON EXTRACT

THE important ingredients of lemon extract are lemon oil and citral, its aromatic constituent. Oil of citronella and oil of lemon grass are sometimes substituted for lemon oil. Methyl alcohol is sometimes used in place of the more expensive spirit alcohol as a solvent for the lemon oil.

LEMON OIL

The presence of lemon oil may be detected by adding a large excess of water to a small amount of the extract in a test tube. If the mixture does not show some cloudiness, it is a strong indication that no lemon oil is present. The degree of cloudiness gives an idea of the amount of oil present.

CITRAL

This is present in the oil of lemon grass, which is sometimes used as a substitute for lemon oil. Citral may be detected by the following test by Burgess:

Add 20 cc. of sulfuric acid to 85 cc. of water. Dissolve in this mixture 10 grams of mercuric sulfate. Shake 2 cc. of the sample with 5 cc. of this reagent in a test tube. If citral is present, the liquid will be bright red, and will quickly disappear and give place to a whitish compound, which floats on top.

Oil of Citronella

This is often substituted for lemon oil. It may be detected by the same test which was used for citral. Instead of the red color and the white compound, citronella gives a bright yellow color, which does not disappear for some time.

Tartaric or Citric Acid

Precipitate the oil by the addition of an equal volume of water. Filter and add a very little of the filtrate to a test tube nearly full of cold lime water. A precipitate will form if tartaric acid is present and it will dissolve in an excess of ammonium chlorid or acetic acid. Filter, or, in case no precipitate forms, heat the liquid. Citric acid is precipitated in the presence of a large excess of hot lime water.

Methyl Alcohol

Mullikin and Scudder.—Take 2 ft. of No. 12-15 copper wire and bend at right angles about 8 or 10 inches from one end. Grasp this bent end and an ordinary lead pencil side by side in such a way that the bend will be about the middle of the pencil. Wind the wire around the pencil and toward the free end of the short part of the wire until you have a coil 3 or 4 cm. long. Remove the pencil and twist the unwound parts together for a handle for the coil.

Dilute a portion of the sample 3 or 4 times, and oxidize 10 cc. of the diluted liquid (preferably in a test tube) by heating to a red heat the above copper coil in the oxidizing flame of a Bunsen lamp. Thrust the heated coil quickly into the liquid contained in the test tube. In a second withdraw and immerse in water. Continue this operation till the oxid of copper fails to be reduced (usually 4 or 5 times is sufficient). Cool the liquid by immersing the tube in water. Separate into two parts and test each for formaldehyde by the following methods:

Mullikin, S. P.—Place one of these parts in an evaporating dish, and add to it 1 cc. of strong ammonia, boil gently over the free flame till the vapors cease to smell of ammonia. Add 2 or 3 drops of strong hydrochloric acid and heat just to boiling, and cool quickly by dipping the dish in cold water. Make the test for formaldehyde: Add a drop of a solution of resorcin (1 : 200) and pour this mixture slowly down the side of an inclined test tube which contains 3 cc. of strong sulfuric acid, taking care to keep the liquids separate. After 3 minutes give the tube a rotary motion by rolling between the hands for a minute or more but only gradually mixing the water and acid, but mixing only about half of the acid.

Flecks of a rose-red color form if methyl alcohol is present. Bands of color or flecks of other colors, even though they be tinged with red or a rose-red solution without the precipitate, should never be taken as proof that methyl alcohol is present. These conditions, however, are good grounds for repeating the test; 10 per cent or even less may be detected by this test.

Hydrochloric Acid and Ferric Chlorid Test.—Add a few drops of the other part of the above oxidized liquid to about 10 cc. of milk, known to be free from formaldehyde, in a porcelain casserole, and add 10 cc. of commercial hydrochloric acid (sp. gr. 1.2) which contains 1 cc. of 10 per cent ferric chlorid per 500 cc. Heat slowly over the open flame nearly to boiling. Give the liquid a rotary motion to break up the curd. If formaldehyde is present, the liquid will be colored violet. If not, it slowly turns brown. The presence of formaldehyde proves that methyl alcohol was in the original extract.

Coloring Matter

Preliminary Test.—Treat the sample with strong hydrochloric acid, and if tropæolin or methyl orange be present the solution will turn pink; Martius yellow partially decolorizes the solution; dinitrocresols decolorizes the solution. Turmeric or naphthol yellow produces no color change.

Turmeric

Turmeric may be detected by soaking a piece of filter paper in the sample, drying and dipping it in a dilute solution of boric acid or borax which has been slightly acidulated with hydrochlorid acid. Dry again and a cherry-red color forms if turmeric is present. Add a drop of dilute alkali and if turmeric be present the paper will be colored dark olive.

Coal-Tar Colors

Evaporate some of the extract to dryness; take up the residue with water and extract the coal-tar colors if present, and test for them by the method given under canned vegetables.

VANILLA EXTRACTS

The best grades of vanilla extract are made by treating vanilla beans with 50 per cent alcohol. Coumarin, an extract from tonka beans, may be used in making the extract. This of course would make a cheaper product. If less than 50 per cent alcohol is used in making the extract, some alkali must be added to dissolve the resins which will not dissolve in a weaker alcohol. In artificial extracts some such coloring matter as caramel or tannin is used.

Preliminary Test.—To a portion of the extract add a few drops of lead acetate solution. The absence of a bulky flocculent precipitate shows the extract not to be of high quality. Leach recommends that normal acetate of lead be added to the sample, and if a precipitate does not form it is conclusive evidence that it is not a pure extract.

When a precipitate forms with the above reagent, it should settle immediately and leave a clear and almost colorless liquid. When there is a mere cloudiness only, it may be due to caramel, in which case the extract is to be suspected.

Alkali

Shake a portion of the sample with twice its volume of water. If no precipitate forms, an alkali is present. A flocculent reddish-brown precipitate shows no alkali is present. If the solution is milky it indicates the presence of a foreign resin.

Add hydrochloric acid drop by drop to the diluted extract. Nothing more than a mere turbidity should result. Should it be quite turbid and the color fading after a time, it shows that an alkali has been used.

Foreign Resins

Mix a portion of the extract slowly with twice its volume of water, frequently shaking the mixture. When this solution is milky, it indicates a foreign resin.

Hess' Test.—Dealcoholize 25 cc. of the sample by concentrating on the water-bath, adding water from time to time to retain the original volume. When no alkali is present in the extract, pure vanilla resin will be thrown down as a reddish-brown flocculent precipitate. Collect the resin, whatever its color, on a filter, and wash. Save the filtrate to test for caramel. Place a piece of the paper and resin in a dilute solution of potassium hydroxid. If the resin is that of pure vanilla it will dissolve, giving a deep-red color, and is reprecipitated when the alkali is neutralized with hydrochloric acid. Dissolve another part of the precipitate in alcohol, and to a part of this solution add a few drops of ferric chlorid; and to the other part, hydrochloric acid. There should be no marked coloration in either case if the resin is that of pure vanilla. Foreign resins nearly always produce a coloration.

CARAMEL

Shake the bottle of vanilla, and if the bubbles, which form, are a bright caramel color, keeping the color till all are gone, the presence of caramel is indicated.

Concentrate a portion of the filtrate, which was saved in making the test for foreign resins, at a rather low temperature until it has about the same color as the original extract. Add a few drops of strong hydrochloric acid and heat very gently. If caramel is present, a yellowish-red flocculent precipitate will form. After the liquid cools, filter and wash with water. Should this precipitate contain caramel, it will not dissolve in water, ether, and alcohol, but will dissolve in sodium hydroxid, dilute alcohol, and glacial acetic acid.

Tannin

Test another portion of the filtrate made in testing for foreign resins, with a few drops of a solution of gelatin. A slight precipitate only should form due to the presence of a small amount of tannin normally present in this filtrate. A large excess shows that it has been added to the extract.

Coumarin

Leach's Test.—Dealcoholize a portion of the sample as above and treat with ammonia, add 3 or 4 volumes of chloroform in a separatory funnel. Evaporate the chloroform extract in an oven, not permitting the temperature to rise above 60° C. To the residue add a few drops of water; warm gently, and add a little of a solution of 1 gram of crystallized potassium iodid in 50 cc. of water, and the solution saturated with iodin. If coumarin is present, a brown, precipitate will form, and if stirred with a rod it will collect in dark green flecks.

CHAPTER VIII

SACCHARINE PRODUCTS

Honey—Adulterations of—General observations—Cane sugar—Commercial glucose syrup—Gelatin.

Maple Syrup—Adulterations of—General examination—Glucose.

HONEY

BEES are sometimes fed with cane sugar. Often glucose syrup is poured over honeycomb from which the honey has been extracted, and the mixture sold as genuine honey.

Gelatin may be added to increase the weight or to thicken the more voluble adulterants.

The ash of genuine honey is not over 0.3 per cent. Whenever the ash is greater than this it should be tested for calcium sulfate, the presence of a considerable quantity of which is an almost certain proof that starch glucose or invert sugar has been added to the honey. Sulfates may be detected by adding barium chlorid to the aqueous solution of the honey and precipitating barium sulfate.

If the ash is high and considerable chlorids are present, molasses has quite probably been added to the honey. The presence of chlorids may be determined by the addition of silver nitrate which precipitates silver chlorid.

CANE SUGAR

The presence of cane sugar can be detected with certainty only by the use of the polarimeter. Its presence in large quantity gives a high right-handed rotation.

COMMERCIAL GLUCOSE SYRUP

Allen's Test.—Make the test for dextrine which is present in commercial glucose, but not in pure honey. Dilute a portion of the honey with an equal volume of water and add methyl alcohol with constant stirring until there is a permanent turbidity. If glucose syrup is present a heavy gummy precipitate will soon form. Genuine honey gives only a slight milkiness.

GELATIN

Dilute a portion of the sample and add a solution of tannic acid. A precipitate indicates the presence of gelatin.

Treat the sample with alcohol, and gelatin, if present, will be left undissolved, and it will give its characteristic odor on ignition.

MAPLE SYRUP

This is sometimes adulterated with glucose, molasses, golden syrup, and with ordinary white sugar. There are no satisfactory simple chemical tests for these substances.

Pure maple syrup should have an ash not lower than 0.35 to 0.40 per cent. A lower ash shows that cane sugar has been added. A higher ash would indicate the presence of molasses or brown sugar stock. These last two adulterants, if present in great abundance, may be detected by taste.

Glucose

This may be detected by the use of the polarimeter. Pure maple syrup gives 53.1 to 60 direct, and—22.2 to—21.9 after hydrolysis. Maple syrup adulterated with glucose gives 80 to 100 direct and 18.9 to 45.6 after hydrolysis (according to Ogdon).

CHAPTER IX

SPICES

Mustard—Adulterations of—Flour—Coloring matter—Turmeric—Martius yellow or analogous coal-tar coloring matter—Cayenne pepper.

Pepper—Adulterations of—General test—Ground olive stones—Cayenne pepper.

MUSTARD

MUSTARD is often adulterated with mustard hulls, wheat, and rice. And when white-colored flour of any kind is used, turmeric, Martius yellow, or a coal-tar color is employed to give the mixture the color of mustard. Cayenne pepper is occasionally used to impart pungency to diluted mustard.

FLOUR

Boil 2 grams of the mustard in 4 or 5 cc. of distilled water for about 10 minutes. After it is cool, add a few drops of iodin solution slowly, avoiding a large excess though having a little uncombined iodin. If a blue color is produced, some starchy matter has been added to the mustard. The intensity of the reaction is an indication of the amount of starchy matter used. Pure mustard contains no starch and hence gives no reaction with iodin.

COLORING MATTER

Pure mustard is a very light dull yellow, and whenever the sample is bright yellow, there is good grounds for suspecting the presence of some artificial coloring matter.

TURMERIC

Add strong ammonium hydroxid to the mustard, and if turmeric is present an orange-red color is usually produced.

Make an alcoholic extract of the sample and dip a piece of filter paper in it, and when dry draw it through a cold, saturated solution of boric acid in water. An orange or red-brown tint produced on the paper indicates the presence of turmeric.

Thoroughly mix 2 or 3 grams of the mustard with castor oil and filter. If turmeric is present the filtrate will appear fluorescent.

Extract a portion of the sample with 3 times its weight of wood alcohol and filter. Evaporate one half of the solution to dryness and add a little hydrochloric acid to the residue. This will turn red whenever turmeric is present, and if an excess of alkali be added it will change to a greenish blue. Evaporate the other half to dryness and moisten with a solution of boric acid and dry on a steam bath. A cherry-red color indicates turmeric.

MARTIUS YELLOW OR ANALOGOUS COAL-TAR COLORING MATTER

Extract the slightly acidified sample with 95 per cent alcohol and dye wool as directed under "Vegetables." The wool will be dyed a bright yellow.

Allen's Test.—Treat a portion of the sample with cold alcohol, and shake vigorously for 5 minutes, then filter and evaporate the filtrate to dryness; add enough water to take up the residue and dye some white wool in this liquid as in the last test. When the dyed wool is wrapped in white paper and heated to 120° in an air bath, part of the coloring matter will be transferred to the paper. The coloring matter dissolves readily in dilute ammonia or hot water, and on the addition of hydrochloric acid the solution is decolorized and a yellow precipitate formed. This distinguishes it from picric acid.

Cayenne Pepper

Allen's Test.—Boil 1 gram of the mustard for a few minutes with alcohol, filter, and evaporate to dryness at about 100°. Taste the residue and cayenne may be recognized by its pungency. Or heat a portion of the extract, and smell the fumes. Irritation of the lungs and coughing will surely follow if cayenne pepper is present.

PEPPER

Pepper may be adulterated with wheat, buckwheat, pepper husks, ground olive stones, spent ginger. Cayenne pepper is sometimes added to adulterated pepper to give it the normal pungency. Many of these adulterants can be detected only by the aid of the microscope.

Neuss's Test.—True pepper turns an intense yellow when covered with strong hydrochloric acid. Any adulteration can be detected at once by the color.

Ground Olive Stones or "Poivrette"

Make a paste of the pepper with caustic alkali. Dilute with a large quantity of water and wash by decantation. Olive stones will be colored a bright yellow; pepper-husks will appear dark.

Jumeau's Test.—Dissolve 5 grams of iodin in a mixture of 50 cc. of ether and 50 cc. of alcohol. Cover the bottom of a porcelain capsule with the finely ground pepper, and add just enough of the iodin mixture to wet the entire mass, and mix well till it has the same consistency throughout. Let dry in the air, then powder and examine it, and if olive stones are present they will be colored yellow. Pure pepper would have a deep brown color.

Aniline acetate, one part aniline in 3 parts acetic acid, colors pure pepper gray or white and olive stones yellowish brown.

Cayenne

Heat some of the red particles found in the pepper and their characteristic vapor is produced. Dissolve the particles in alcohol or ether and the same vapors are produced.

CHAPTER X

VINEGAR

Adulterations of—Preparation of sample—General observations—Free mineral acids—General tests—Sulfuric acid—Hydrochloric acid (free)—Malic acid—Coloring matter—Caramel—Coal-tar colors—In wine vinegar—Free tartaric acid—In wine vinegar.

VINEGARS may be adulterated by the addition of mineral acids as sulfuric or hydrochloric. Caramel or the coal-tar dyes may be employed to improve the color or to give color to an artificial product. Malic acid is always present in cider vinegar. Potassium acid tartrate occurs in true wine vinegar. Poisonous metals may be present in vinegars containing free mineral acid. Entirely artificial cider vinegar is often found on the market.

Preparation of the Sample for Testing

If the vinegar is turbid from any suspended matter, it should be filtered. The samples should be analyzed at once, and in the laboratory they should always be kept in glass-stoppered bottles.

General Observations.—Ignite a little of the vinegar residue on a clean platinum wire in a colorless Bunsen flame, and if it is pure cider vinegar the flame will be colored the characteristic lilac color of potassium. The sodium flame is absent or only a mere trace of it is present. But in all artificially colored vinegars, spirit sugar and glucose vinegars, the sodium flame predominates.

The residue of cider vinegar is thick, viscid, or mucilaginous, of a light brown color, astringent acid taste though not unpleasant. The solids of sugar-house vinegar, those from colored spirit and wood vinegar, each have a bitter taste on account of the caramel used to color them. The residue of the sugar-house vinegar has the odor of molasses. Wood vinegar when present gives a residue with a tarry or smoky taste and smell. Glucose vinegar gives the odor of scorched corn. Solids of fruit vinegars are quite soluble in alcohol, except a granular residue in grape vinegar, while the solids of malt and glucose vinegars are almost insoluble.

The ash of fruit vinegars and malt vinegars has a distinct alkaline reaction, while that of spirit and wood vinegars is very feebly alkaline.

FREE MINERAL ACIDS

The ash of pure cider vinegar is always alkaline. If a vinegar should show a neutral reaction this would certainly indicate the presence of a free mineral acid. If the ash be alkaline, no acid except nitric could have been present, and this is seldom, if ever, used as an adulterant of vinegar.

When the Ash is Alkaline Apply

Ashby's Test.—Extract 0.5 gram of logwood in 100 cc. of water and dry a drop or two on a porcelain surface. Then add a drop of the vinegar and dry again. If the residue is red, a mineral acid is present; if yellow, mineral acids are absent. When only a very small amount of the acid is present the red coloration will be destroyed on diluting with water, but may be restored by concentrating the liquid.

Sulfuric Acid

Sulfuric acid, if present, will cause the vinegar to leave a charred mass when evaporated over the water-bath.

Frear's Method.—Mix 5 cc. of the sample and 5 or 10 cc. of water, and add a very little of a solution of methyl violet (made by dissolving one part of methyl violet 2 B. in 100,000 parts of water). A blue or green coloration shows the presence of mineral acids.

Sulfuric Acid as Distinguished from Sulfates

Allen's Method.—Evaporate 100 cc. of the vinegar down to one tenth its volume, and when cold add 50 cc. of alcohol. Sulfuric acid remains in solution while the sulfates are precipitated. Dilute the solution and precipitate the acid with barium chlorid.

HYDROCHLORIC ACID

Free

Place a definite quantity of the vinegar in a distilling flask and distil off half. Add a few drops of silver nitrate to the distillate. If a precipitate forms, hydrochloric acid is present.

MALIC ACID

Leach's Method.—To 5 cc. of the sample, add a few drops of a solution of calcium chlorid (1 : 10); make slightly alkaline with ammonia. Filter off any precipitate that may form, add 20 to 30 cc. of 95 per cent alcohol to the filtrate and heat to boiling. If malic acid is present, a voluminous flocculent precipitate will form. A precipitate may form in vinegars containing dextrine. Make a further test for malic acid by the following: Filter and treat the precipitate with a little alcohol, and when dry add concentrated nitric acid and evaporate to dryness on a water-bath. Treat the residue with sodium carbonate, boil for a short time, filter. Add acetic acid to the filtrate till slightly alkaline, boil till carbon dioxid is expelled, and if on the addition of calcium sulfate a precipitate forms, it indicates the presence of malic acid.

COLORING MATTER

Caramel

The residue of vinegar to which much caramel has been added has an unusually dark color and bitter taste.

Crampton and Simons' Method.—Shake well together in a corked flask 50 cc. of the vinegar with about half as many grams of fullers' earth; after standing for half an hour filter. Vinegar containing no artificial color will show scarcely any change in color when thus treated. A caramel-colored vinegar will be decolorized in proportion to the amount of caramel present.

Coal-Tar Colors in Wine Vinegar

Test by the usual test for coal-tar dyes. See under canned vegetables.

METALLIC IMPURITIES

Vinegars containing free mineral acids are sometimes found to contain poisonous metals.

Evaporate 200 to 400 cc. of the vinegar to dryness, add a little sodium hydroxid to this residue and burn to an ash over a low flame. It may be necessary to add a little potassium nitrate once or twice. Add a little dilute hydrochloric acid and saturate with hydrogen sulfid and test for lead, zinc, copper, and arsenic according to Allen's method given under canned meats.

Spices to Increase Pungency

Leach.—Neutralize a portion of the vinegar with sodium carbonate. The presence of spices is easily detected by tasting this mixture.

Another Test.—Exactly neutralize a little of the vinegar as above, evaporate to smaller bulk and taste as before, then shake the concentrated liquid with ether, separate the ethereal layer and evaporate it, and taste the residue.

Tartar in Wine Vinegar

The presence of tartar in vinegar proves it to be wine vinegar.

Allen's Method.—Evaporate a portion of the vinegar and treat the residue with alcohol; a granular residue of tartar remains undissolved. To prove that it is tartar, decant the alcohol and dissolve the residue in a little hot water, cool, rub the inside of the vessel with a glass rod, and if tartar was present acid potassium tartrate will be deposited where the rod touched the vessel. The test will be more sensitive if an equal volume of alcohol is added.

Free Tartaric Acid in Wine Vinegar

Test as for Tartar.—Treat the alcoholic solution of the extract with an alcoholic solution of potassium acetate. Rub the sides of the vessel as before, and if tartaric acid is present the streaks and sometimes a precipitate forms where the rod touches the vessel.

GLUCOSE

Glucose is present when both direct and invert readings are dextro-rotatory.

CHAPTER XI

FATS AND OILS

Lard—Adulterations of—Cottonseed oil—Cottonseed stearin—Beef stearin.

Olive Oil—Adulterations of—General test—Cottonseed oil—Peanut oil—Sesame oil—Rape oil.

LARD

LARD is very often adulterated with cottonseed oil, cottonseed stearin and beef stearin. Their being very much cheaper accounts for the sophistication.

COTTONSEED OIL

Halphen's Test.—Dissolve 1 per cent of sulfur in a given volume of carbon bisulfid. Add an equal volume of amylic alcohol. Mix 3 to 5 cc. of this reagent with an equal volume of the melted lard in a test tube. Close with a cotton stopper and boil for 15 minutes in a bath of saturated brine. The presence of cottonseed oil is indicated by a deep-red or orange color, little or no color resulting in its absence. Lard from hogs fed on any of the various cottonseed products may give a faint reaction when this test is applied.

COTTONSEED STEARIN

Since cottonseed stearin is only the more solid portions of cottonseed oil, the above test may be applied, but to distinguish it from the latter it is necessary to make determinations quite beyond the scope of this set of tests.

Beef-Stearin

It is very difficult to identify beef-stearin by chemical tests. It is usually detected by use of the microscope. Leach gives the following method: Make a solution of 2 to 5 grams of the fat in 10 to 20 cc. of ether. Let stand a half day, at about the room temperature. Loosely stopper the tube with cotton to prevent too rapid evaporation of the ether. It is well to vary the conditions of heat, amount of solvent, and rate of crystallization, to get the best possible results. It may often be well to separate the crystals thus obtained by filtering and recrystallizing from ether. Separate the crystals that form at the bottom of the test tube from the liquid portion by pouring on a small filter. Wash them several times with ether, but not sufficient to remove the mother liquor entirely. In case it is all removed, and the crystals are too fragile to mount, add a drop of alcohol. Crystals of lard stearin are flat rhomboidal plates, one end being oblique to the sides, and they do not appear to be regularly grouped. Beef-stearin crystals are rod-shaped, or needles often apparently curved with pointed ends, and are arranged in clusters like the ribs of a fan, the crystals radiating from a common point. Under certain conditions the lard crystals are not irregularly grouped, but are arranged like the parts of a feather, where one part seems attached to another close at hand. Considerable experience is necessary to use this test with absolute certainty.

OLIVE OIL

Olive oil is one of the most commonly adulterated foods. The commonest adulterant probably is cottonseed oil. Other foreign oils, such as peanut, sesame, and rape, are sometimes used.

Preliminary Test.—Pure olive oil turns from a pale to a dark-green color in a few minutes, when it is shaken with the same volume of concentrated nitric acid or sulfuric acid. Whenever a reddish to an orange, or brown coloration results, the presence of a foreign vegetable oil is indicated (probably a seed oil).

Bach gives the following results of strong nitric acid on the common oils. Olive oil when shaken with nitric acid gives a pale green, which changes to an orange yellow after heating five minutes. With similar treatment peanut oil gives pale rose and brownish yellow; rape oil, pale rose and orange yellow; sesame oil, white and brownish yellow; sunflower oil, dirty white and reddish yellow; cottonseed oil, yellowish brown and reddish brown; castor oil, pale rose and golden yellow.

Pontet's Test; Elaiden Test.—Treat 1 cc. of mercury with 12 cc. of cold nitric acid (sp. gr. 1.42) and shake 2 cc. of this freshly-made solution with 50 cc. of the sample in a bottle every 10 minutes for 2 hours. Oils which are principally olein, or mixtures of olein and solid esters like palmatin and stearin, give more or less solid products, but olive oil is remarkable for the firmness of the canary or lemon-yellow mass which is formed. After standing a day the mass cannot be pierced with a glass rod and sometimes it gives forth a sound when struck.

This test requires considerable experience to be used with any great degree of certainty.

Cottonseed Oil

Carbon bisulfid containing 1 per cent of sulfur in solution is mixed with an equal amount of amyl alcohol. Equal volumes (about 3 cc.) of this reagent and the sample, are mixed in a test tube, which is loosely stoppered with cotton, and heated in a bath of boiling saturated brine for a quarter of an hour. The presence of cottonseed oil is shown by the formation of a deep-red or orange color. Little if any color is produced in its absence. If no color is produced it is well to add another cc. of the reagent and heat 5 or 10 minutes more, and to repeat this again if no color forms. Lard and lard oil from animals fed on cottonseed meal may give a faint reaction.

Peanut Oil (Arachis Oil)

Bellier's Test.—Saponify a gram of the sample with 5 cc. of a solution of 85 grams potassium hydroxid in a liter of strong alcohol. This may be done in a small Erlenmeyer flask on the water-bath. Then boil for two minutes, neutralize exactly with dilute acetic acid (use phenolphthalein as the indicator). Cool the mixture by placing the flask in water at 17° to 19° C. A precipitate usually forms. Add 50 cc. of 70 per cent alcohol which contains one per cent by volume of concentrated hydrochloric acid (sp. gr. 1.2). Shake the flask vigorously and cool again as before. If no precipitate forms the oil is not adulterated with peanut oil. The presence of 10 per cent or more of peanut oil produces a precipitate, even a smaller amount will produce a cloudiness after standing between 17° and 19° C. for 30 minutes.

Some varieties of olive oil from Tunis give the same turbidity when the 70 per cent alcohol is added. To distinguish these from peanut oil heat the mixture on the water-bath till everything has dissolved, and cool to 17° to 19°. The cloudiness will not appear if the oil is pure, but will reappear if peanut oil is present.

Sesame Oil

Badouin's Test.—About 0.1 of a gram of cane sugar is dissolved in 10 cc. of hydrochloric acid (sp. gr. 1.20) shaken vigorously with 20 grams of the sample for a minute or more. After standing for a while the aqueous solution will separate from the oil. If 1 per cent or more of sesame oil is present, the aqueous solution will be colored crimson.

Tocher's Test.—Dissolve 1 gram of pyrogallic acid in 15 cc. of strong hydrochloric acid. Add an equal volume of the oil in a separatory funnel. When it has stood a minute, draw off the aqueous solution and boil. In the presence of sesame oil it is colored red by transmitted light, and blue by reflected light.

Rape Oil

Palas' Test.—Make a 1 per cent solution of fuchsin and a 30 per cent solution of sodium acid sulfite. Mix together 20 cc. of each of these solutions and add 200 cc. of water and 5 cc. of strong sulfuric acid. After the solution is decolorized, 10 cc. of the sample is shaken with it. If rape oil is present, the color will be partially restored. To prevent the formation of the color by contact with the air have the vessel full of the mixture.

CHAPTER XII

BEVERAGES

Coffee—Adulterations of—General test—Coloring matter—Imitation coffee beans—Chicory.

Tea—Adulterations of—Foreign leaves—Exhausted tea leaves—Lie tea—Facing—Catechu.

COFFEE

COFFEE is often colored with such substances as Scheele's green, chrome yellow, iron oxide, Prussian blue, indigo and turmeric. Imitation coffee beans have been made of wheat flour, bran, rye, chicory and peas.

Allen's Preliminary Test.—A good preliminary test for ground coffee is to sprinkle some of it on the surface of cold water. The oil of true coffee prevents the particles from being readily soaked, and so they float for some time. Chicory and most of the other adulterants of coffee contain no oil, but do contain caramel, which is quickly extracted by the water producing a zone of brown color about such particles. They become soaked and quickly sink. The liquid containing pure coffee diffuses uniformly without coloring the water to any perceptible degree. Chicory and similar roots give a dark brown, turbid infusion. Roasted cereals do not impart so distinct a color to water.

Coloring Matter

Shake the coffee beans in cold water and make the regular qualitative tests for the inorganic coloring matters—Scheele's green may be identified by testing for copper and arsenic; chrome yellow, by testing for lead chromate; iron oxide may be detected by its characteristic tests.

Organic coloring matter is best extracted with alcohol. Prussian blue may be detected by dissolving it from the sediment with hot caustic alkali, acidifying with hydrochloric acid, treating it with a drop of ferric chlorid. If present, ferric ferrocyanide, a blue precipitate, will be formed. Indigo is not discharged by sodium hydroxid, while Prussian-blue is. It will form a deep blue solution with sulfuric acid.

Test for turmeric as under mustard.

Imitation Coffee Beans

Most imitation coffee, as already stated, is heavier than water. Coffee contains no starch, so the imitation beans made of cereals may be detected by testing for starch.

Starch

Allen's Method.—Boil the coffee in 10 parts of water. When perfectly cold add to it a little sulfuric acid, then a strong solution of potassium permanganate, drop by drop, with constant shaking, till the liquid is almost decolorized; strain or decant and add to the solution a solution of iodin. If 1 per cent or more of starch is present, a blue coloration will be produced.

Chicory

Rimmington's Test.—Boil a portion of the sample with water which contains a little sodium carbonate; decant, wash and treat the residue with a weak solution of bleaching powder for several hours. The solution will be decolorized. The coffee will be at the bottom as a dark layer while the chicory will be a light layer above it.

Albert Smith's Test.—Boil 10 grams of the sample in 250 cc. of water; strain and add basic lead acetate in slight excess. A precipitate forms, and when it has settled the supernatant liquid will be colorless if the coffee is pure, but more or less colored if chicory is present.

TEA

Tea is adulterated by the substitution of inferior grades for those of better quality, by the addition of exhausted leaves and foreign leaves, by the use of coloring matter or "facing" such as Prussian blue, indigo, or turmeric to color green tea, and sometimes graphite to color black tea. Foreign astringents (generally catechu) are added to conceal the presence of exhausted leaves. An imitation tea, "lie tea," is made of the stems and dust with mineral matter, and some starch or gum to hold these together.

Foreign Leaves

Though there are several chemical tests for foreign leaves, none are as satisfactory as a microscopical examination. Soften the leaves by soaking in hot water, unroll carefully and examine with a hand lens or low power of the microscope. Compare with a genuine leaf—the shape, margin, and venation.

Exhausted Tea Leaves

Sometimes such leaves may be detected by a physical examination. They are often more or less unrolled and broken on the edges. But the only certain way of ascertaining their presence is to determine the soluble ash which is from 2.5 to 4 per cent in pure tea and usually less than 0.8 per cent in exhausted tea.

Lie Tea

This imitation tea is easily detected by pouring hot water over the leaves.

If they are artificial, they will break down into the fragments of which they were made.

Facing

Organic coloring matter may be detected by the same method used for detecting such colors in coffee.

Catechu

Hager's Test.—Boil a little of the tea in water, and add to the extract an excess of lead monoxid. If the tea is pure the addition of a solution of silver nitrate produces only a slight grayish precipitate, but when catechu is present a yellow flocculent precipitate forms.

A FEW OF THE BEST BOOKS ON FOOD ANALYSIS

ALLEN, A. H., Commercial Organic Analysis, 1898. Price, $29.50.

BLYTH, A. W., Foods, their Composition and Analysis, 1903. Price, $8.40.

HASSALL, A. H., Food, its Adulterations and the Methods for their Detection, 1874.

LEACH, A. E., Food, Inspection and Analysis, 1905. Price, $7.50.

LEFFMANN, H., and BEAM, W., Select Methods of Food Analysis, 1905. Price, $2.50.

U. S. Dept. of Agr., Bureau of Chem., Bulletin 13, Parts 1-10, Food and Food Adulterants, 1887-1902. (Some of these are out of print.)

Bulletin 46, Methods of Analysis of Foods, 1899; Bulletin 65, Provisional Methods for the Analysis of Foods, 1902.

WILEY, H. W., Foods and their Adulteration, 1907. Price, $4.00.

CHEMICALS

The following are the chemicals used in making all the tests in this book. Most of these are found in every chemical laboratory.[1]

Acetic acid

Ammonium carbonate

Ammonium chlorid

Ammonium hydroxid

Ammonium sulfid

Amyl alcohol

Aniline

Barium chlorid

Barium peroxide

Boric acid

Bromine

Calcium carbonate

Calcium chlorid

Calcium sulfate

Cane sugar

Carbon bisulfid

Castor oil

Chloroform

Diphenylamine

Ether

Ethyl alcohol

Ferric chlorid

Formalin

Fuchsin

Gelatin

Glycerol

Hydrochloric acid

Iodin

[1] The more uncommon ones may be obtained from Eimer and Amend, 205-211 Third Avenue, New York.

Lead acetate (basic)
Lead oxide
Lime water
Litmus paper
Logwood chips
Mercury (metallic)
Mercury sulfate
Methyl alcohol
Molybdic oxide
Nitric acid (sp. gr. 1.42)
Oxalic acid
Petroleum spirit
Phenolphthalein
Phenylhydrazine hydrochlorid
Phosphoric acid
Potassium bichromate
Potassium carbonate
Potassium chromate
Potassium ferrocyanid
Potassium hydroxid
Potassium iodate
Potassium iodid
Potassium nitrate
Potassium permanganate
Pyrogallic acid
Resorcin
Silver nitrate
Sodium acetate
Sodium acid sulfite
Sodium carbonate
Sodium chlorid
Sodium hydroxid
Sulfur
Sulfuric acid
Tannic acid
Turmeric paper
Zinc (arsenic free)

INDEX OF AUTHORS AND TESTS

INDEX

A

B

C

G

H

K

L

M

N

O

P

R

S

V

W

Z

9 789361 380112

Printed by Libri Plureos GmbH in Hamburg,
Germany